YOUR KNOWLEDGE HAS VALUE

- We will publish your bachelor's and master's thesis, essays and papers

- Your own eBook and book - sold worldwide in all relevant shops

- Earn money with each sale

Upload your text at www.GRIN.com and publish for free

Pascal Kaufmann

Starmind

Band 3

GRIN Verlag

How many mutations are required to produce a human cancer cell?

Assessment of theoretical models and their experimental support

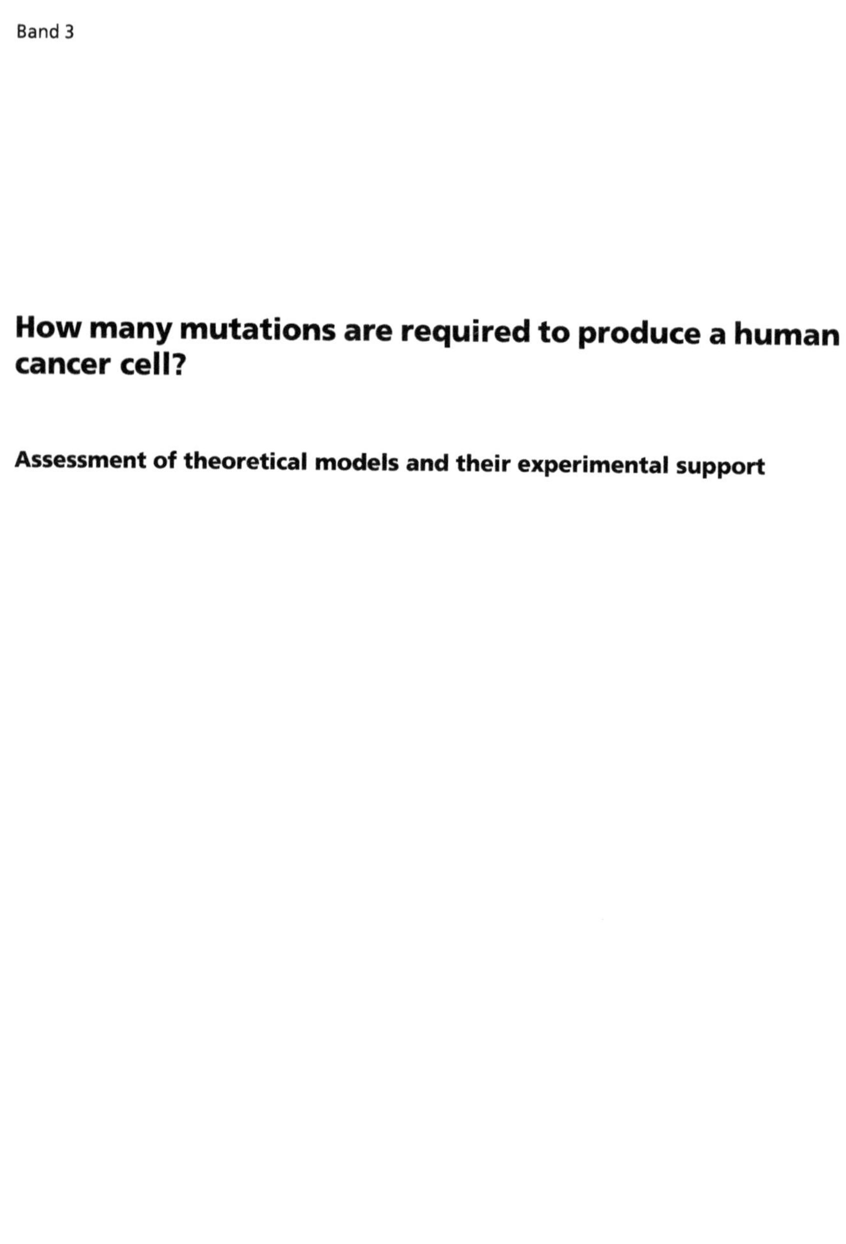

GRIN Verlag

Bibliografische Information der Deutschen Nationalbibliothek:

Die Deutsche Bibliothek verzeichnet diese Publikation in der Deutschen National-bibliografie; detaillierte bibliografische Daten sind im Internet über http://dnb.d-nb.de/ abrufbar.

Imprint:

Copyright © 2011 GRIN Verlag GmbH
Druck und Bindung: Books on Demand GmbH, Norderstedt Germany
ISBN: 978-3-640-85341-0

This book at GRIN:

http://www.grin.com/en/e-book/168293/how-many-mutations-are-required-to-produce-a-human-cancer-cell

Abstract

Data from many diverse fields demonstrates that tumour progression requires the disruption of a distinct set of cellular phenotypes associated with proliferation and survival. It is known that these changes can be brought about by multiple mutations sourced from a wide selection of "cancer genes". It is not known whether a cell can acquire the required number of mutations within the human lifespan solely as a consequence of the background mutation rate. Thus, for some time there has existed a debate regarding the necessity of an increase in mutation rate for tumour progression. It is clear that most cancers do exhibit a mutator phenotype, however it has not been easy to solidify the causal link between it and tumour progression. Here, a wide range of theoretical models and their experimental support are assessed in order to determine which position in the above debate is closer to the truth. Such insights into the fundamental driving forces of cancer hold promise in focusing the search for novel drug targets for its treatment and prevention.

Contents

Introduction

Ordinarily, innate biochemical circuits ensure that physiological control of cellular proliferation is maintained for the benefit of the entire organisms' fitness. Tumour progression involves the stepwise rewiring of these circuits to alter these proliferative controls for the short term fitness benefit of the individual clone. This allows it to outcompete its neighbours in what is accepted to be a process analogous to Darwinian natural selection. There are numerous routes that can be taken to bring about these re-wirings, but as Hanahan and Weinberg (2000) suggest, they are all different manifestations of the same six fundamental hallmarks of cancer (Fig .1.).

It has been suggested that if multiple genetic alterations are required to bring about the above mentioned changes, the low background level of genetic instability may be prohibitive to tumorigenesis. Loeb (1991) postulated that increased genetic instability could account for cancer incidence rates by accelerating the generation of variation. Other authors (Bodmer 2008) have stressed that these models ignore the power of natural selection, which according to his groups models, can account for the genetic changes without increased genetic instability.

However, the multistep Darwinian model alone is too simplistic. As is discussed below, advances in tissue biology suggest the actual population undergoing selection may be limited to a small sub-population at the tip of a cellular hierarchy. If this is the case, it drastically reduces the variability available for selection to act on, potentially increasing the time between the "steps" of tumorigenesis. These observations have even been seen by some (Weinberg 2007) to argue for the addition of an additional hallmark: genetic instability to the six

already suggested in Fig.1. as an accelerating mechanism.

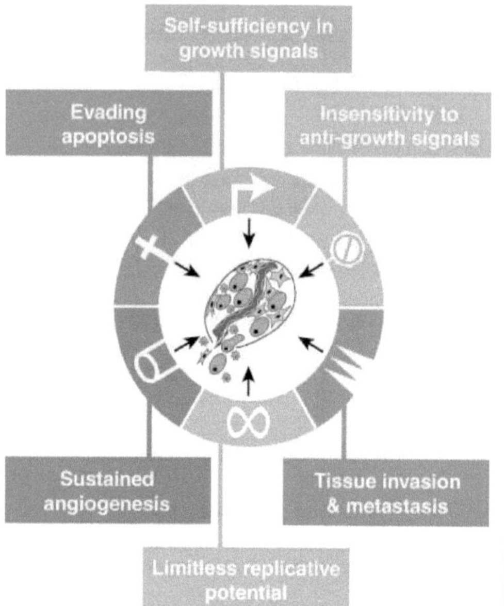

Fig. 1.The six hallmarks of cancer suggested to be necessary for a cell to progress to malignancy together with an example for each pathway:

1. Growth signal autonomy: normal cells rely on a limited supply of growth factors, one of the ways cancers can evade this is to gain 'cellular autonomy' e.g. point mutation of Ras protoncogene results in proliferation stimulated constitutively via the SOS-Ras-Raf-MAPK pathway.

2. Insensitivity to anti-growth signals: e.g. by knock-out or missense mutation of receptors for antigrowth signals such as TGF-b or knockout of downstream signalling molecules e.g. pRb

3. Invasion and Metastasis: this step is important as it accounts for around 90% of cancer deaths and generally involves mutations in or shifts in cell-cell adherence molecules (CAM's) e.g. E cadherin and changes in ECM degrading/remodelling protease expression/secretion.

4. Replicative immortality: 85-90% of tumors have increased telomerase expression

5. Angiogenesis: due to the requirement for nutrients no cell should be further than 100um from a capillary, for the cancer to grow any larger than this it must recruit capillaries. It does this by shifting the balance of the "angiogenic switch" to the proangiogenic side.

6. Evading apoptosis: e.g. mutation of the "guardian of the genome" p53. (Diagram reproduced from Hanahan and Weinberg. 2000)

The plot thickens further with the

advent of the epigenetic era. Epigenetic

alterations may not only offer a complimentary mechanism for generating variability

but potentially an alternative route for initiating and sustaining tumour progression.

Due to the many variables and heterogeneous nature of cancer, no step in

gaining an understanding of it is likely to be straightforward. The question of

whether genetic instability is necessary or not will undoubtedly be the same. Since

much research has been conducted in this area a theoretical approach should

provide some answers to the question of whether genetic instability is necessary.

The approach taken here is to fractionate the problem into its constituent parts and then attempt to find an answer to each individually. When combined, the answers should lead to an overall greater understanding of the question and its answer.

- How many mutations are required to produce a human cancer cell?

- Is genetic instability necessary to acquire sufficient mutations?

- Does raised genetic instability accelerate tumour progression, and is this the mechanism by which human cancer actually evolves?

The theoretical approach can contribute to every step of scientific enquiry involved in answering the question (Fig.2.): in drawing conclusions from the research, in the generation of hypotheses, models, and potentially in generating predictions which can be tested, thus guiding further research. The conclusions reached by this theoretical synthesis should help guide future research into cancer therapies and prevention strategies.

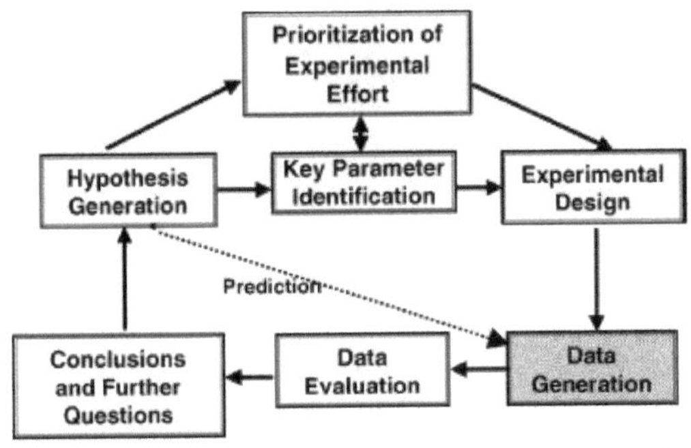

Fig. 2. Role of theoretical approaches in scientific inquiry: Theoretical methods can contribute substantially at every step although it depends on experimental verification. (Reproduced from Beckman and Loeb. 2005)

How many mutations are required to produce a human cancer cell?

A critical parameter which must be considered when trying to assess whether genetic instability is necessary for human cancer is the number of genetic mutations which are necessary to acquire the six hallmarks of cancer described in the last section.

Sequence data

The abundance of data from large scale sequencing has revealed a great deal about the armoury from which cancers choose their arsenal of mutations in the somatic evolutionary arms race. A recent study of a range of cancer types by Greenman *et al.* (2007) revealed 1000 point mutations in the coding regions of 518 genes alone. By looking at the effects of differences in selection pressures, they were able to discern that a surprisingly high number, 120 of the 518 genes were "drivers". However, even when the problem of discerning driver mutations has been overcome, these data only give the average number of mutations which provide a selective advantage. Many of these may be "tweaks" of the genome of the original founding cancer cell which may give the cancer more of a growth advantage, but are not strictly necessary.

This approach is therefore too simplistic. The crucially important question for this discussion is: what is the minimum complement of genetic changes from the above arsenal which is sufficient to produce a cancer cell? Below is presented a summary of the data from several independent lines of investigation which are now converging to come closer to providing an answer to this question.

Epidemiology

Cook et al (1969) and Armitage and Doll (1985) used cancer incidence statistics from a wide range of sources to show that different cancers had a range of best-fit curves with the majority of gradients ranging from 3.5 to 6.5. This made it clear there would be no single answer to the question of how many mutations are required to reach the tumorigenic phenotype and generalising across type (as had been done till then) might be a simplistic way of approaching the problem.

More recently W.D Stein *et al.* (1990), taking cycles of mutation, clonal expansion and selection into account, developed and rigorously tested a simplified version of a two stage model (originally put forward by Moolgavkar and Knudson (1981)) against a wide range of epidemiological data. They argue that in the great majority of cases their data are consistent with at least three mutational events being required to bring about the progression of cancer through the two hypothetical stages.

Thus, the above examples, together with a wider review of the literature (Moolgavkar and Knudson 1981; Lubeck and Moolgavkar 2002), point to between 2 and 7 rate limiting steps leading to a clinically apparent cancer. However this has also highlighted the weaknesses of mathematical analysis of epidemiological data. It is often found that a variety of models can fit the same data, and even if the correct model is found, each rate determining step does not necessarily correspond to the number of mutations.

In vitro data

In vitro transformation assays in the Weinberg lab using viral vectors and oncoproteins have been used to try and distil down the distinct set of changes corresponding to the above steps. Hahn *et al.* (1999) found that disrupting a minimum of five pathways this way: Ras, p53, pRb, PP2A and telomerase could generate tumorigenic cells. Rangarajan and Weinberg (2004) have shown requirements not only vary between murine and human cells but also between different human cells. They found human fibroblasts specifically needed: p53, pRb, PP2A, telomerase, Raf, and Ral-GEFs whereas human embryonic kidney cells required activation of RalGEFs and PI3-kinase but not Raf. These results raise questions about the accuracy of murine models and the practice of attempting to generalise requirements for tumorigenesis across tissue type.

Thus, in Vitro experiments hesitantly point to a minimum of ~five pathways (Weinberg 2007) needing to be disrupted in the majority of cancers which could require five mutations. However, the assay used by Hahn *et al.* only tested for tumorigenic ability to form benign tumours and it is likely more pathways need to be disrupted to progress to full malignancy. The recessive nature of many of the disruptions in the pathways mentioned further increases the uncertainty as to how many mutagenic steps there are in tumorigenesis.

Histopathology

Pathological analyses have long been seen to point to tumour progression occurring

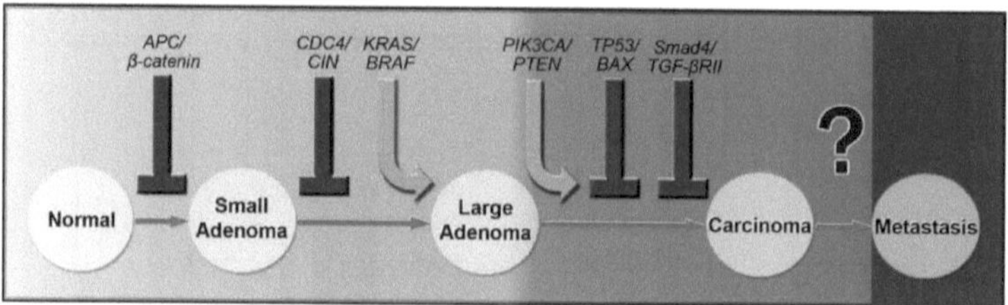

Fig. 4. Major genetic alterations associated with colorectal tumorigenesis. (Adapted from Jones et al. 2008)

as a multistep process (Foulds 1954). Colorectal tumorigenesis has long been known to proceed through a set of well defined clinical stages, each of which is thought to arise due to a characteristic mutation. In a step towards integrating the *in vitro* work with the actual aetiology of cancer, work by Fearon and Vogelstein (1990) resulted in a speculative model of the actual genetic changes occurring during tumour progression. This has been updated and somewhat expanded by Jones et al (2008) using more accurate genetic sequencing (Fig.4.). Such models give some credence to the *in vitro* data and numbers derived thereof above, but it is clear that this sequence is only a single possible route of many and the original data has been seen by some to suggest more steps.

This section has demonstrated that despite large amounts of research having been done in this area, we are still far from making a confident assessment of what the minimum number of genetic mutations is for reprogramming physiologically controlled somatic cells to pathologically deranged cancer cells. Epidemiology suggests between 2 and 7 rate limiting stochastic events. *In vitro* assessment, to some extent corroborated by pathological observation, suggests these could correspond to the disruption of approximately five biochemical pathways, although it is clear that this requirement varies both spatially and temporally in the body. Therefore a commonly used (Hanahan and Weinberg (2000); Rangarajan et al (2004); Beckman and Loeb 2005) number for roughly estimating the number of mutations needed for the majority of cancers is ~6, although as has been shown, this can clearly vary by a great deal.

Is raised genetic instability necessary to acquire sufficient mutations?

The mutator phenotype hypothesis

As described in the last section, tumorigenesis is thought to proceed via a multistep stochastic pathway. It has been suggested by Loeb et al (1991) that the background mutation rate under normal physiological growth conditions may be too low to allow these steps to occur in the fixed timeframe of the human life span. An updated version of the original estimation which led to this proposal is presented here (Beckman and Loeb 2005): The stem cells thought to give rise to tumours have a rate of mutation between 10^{-9} and 10^{-11} per nucleotide locus per division. The maximum number of stem cell generations in a human lifetime is thought not to exceed 10^4 and only a very small fraction of somatic cells are stem cells. Assuming six independent, specific mutations are required to accomplish the six rewiring's mentioned previously, the chances against any stem cell in a human developing sufficient mutations are astronomical. Even when taking the minimum epidemiological estimate of 2 mutations being necessary, the estimated probability of developing a cancer per individual per lifetime would be far less than the levels observed.

To account for this discrepancy they suggested that an accelerating mechanism may be required. They posited that raised genetic instability manifesting as a "mutator phenotype" could boost the rate at which variation is produced

"

thereby shortening the time between tumorigenic steps and allow cancer to develop

at the rates seen.

Arguments which undermine the calculations' assumptions

Many have argued that the assumptions that this argument is based on are flawed.

Each of the six hallmarks need not necessarily be accomplished by mutation

of a specific gene; rather, the biochemical circuit in control of the particular function

must be disrupted. This can be achieved by either over or under expressing any of

the Proto-oncogenes or TSG's respectively at any level of the signalling cascade

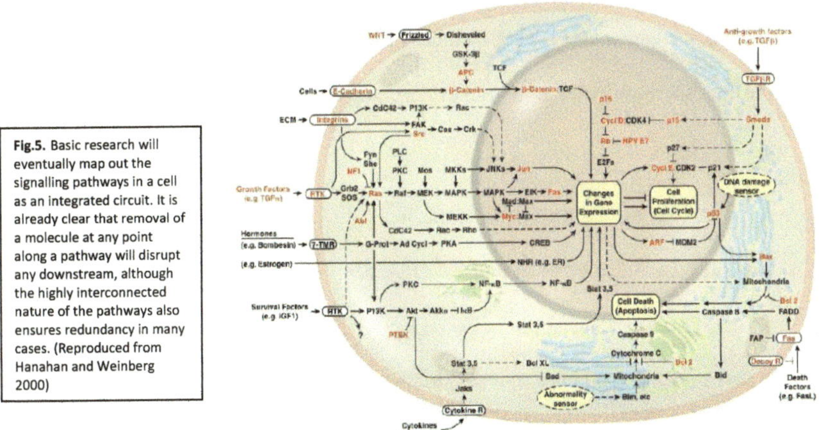

Fig.5. Basic research will eventually map out the signalling pathways in a cell as an integrated circuit. It is already clear that removal of a molecule at any point along a pathway will disrupt any downstream, although the highly interconnected nature of the pathways also ensures redundancy in many cases. (Reproduced from Hanahan and Weinberg 2000)

(Fig.5.), from the ligand right down to transcriptional control, thus increasing the

number of susceptible targets per step.

The assumption that all the mutations are independent i.e. that there is a

ratio of 1 mutation to 1 phenotype is too simplistic. Due to the cyclic and highly

interconnected nature of many biological pathways (Figure.5.), any one mutation is

likely to have several effects. For example a single missense mutation of p53 will

promote multiple phenotypes including evasion of apoptosis and facilitation of angiogenesis. In fact, almost all TSG's and Oncogenes interact indirectly with the angiogenic switch via key controllers such as HIF1 (Vogelstein and Kinzler 2004). Further, mutations may act synergistically, amplifying the effects of a single mutation. A classic example of such an interaction is between p53 and BRCA2. Cells will not tolerate BRCA2 mutations in isolation but show far better growth when p53 is concomitantly mutated (Dong *et al.* 1993).

Clonal evolution and natural selection

As early as 1930 RA Fisher showed that low levels of variation could drive the evolutionary process. Bodmer (2008) has argued that the above argument for genetic instability is ignoring the critically important part played by natural selection in tumorigenesis.

Escape from even a single stringent physiological constraint, may confer a survival advantage to premalignant clones which could allow them to expand in number. This expansion increases the number of replications a cell cycles through, as well as expanding the clonal population, together multiplying the probability of mutation many fold. An example where this can have a profound effect is in the mutation of TSG's, Nowak et al (2004) produced a model of the

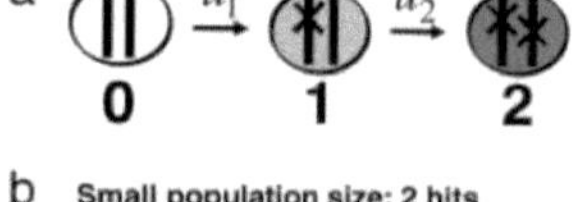

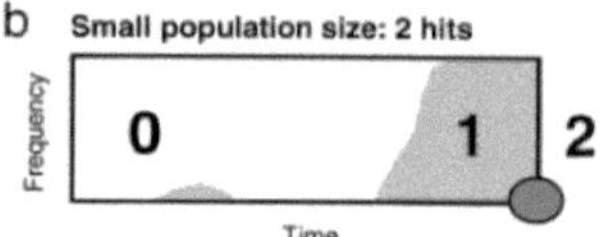

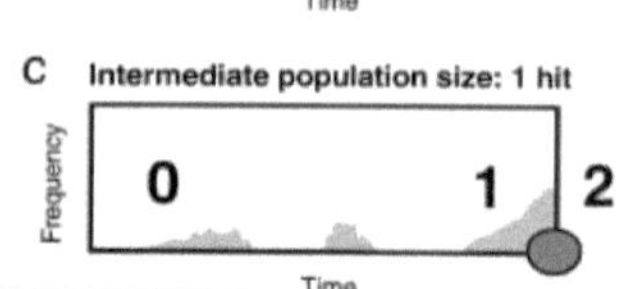

Fig. 6. Inactivating a TSG requires two, one, or zero rate-limiting steps depending on the population size. (*a*) Cells of type 0, 1, and 2 have, respectively, 0, 1, and 2 inactivated alleles of the TSG. The mutation rate for inactivating the first and second allele are given by $u1$ and $u2$. Initially all cells are of type 0. (*b*) In small populations, type 1 cells will reach fixation before a cell of type 2 has been generated. The resulting kinetics have two rate-limiting steps. (*c*) In intermediate populations, a lineage of type 1 cells generates a type 2 cell before reaching fixation. The resulting kinetics have one rate-limiting step. (*d*) In large populations, type 1 cells are generated immediately and accumulates a linear function of time. After some (short) time a type 2 cell will be generated. The kinetics involve two steps, none of which is rate-limiting for overall cancer progression. (Adapted from Nowak et al 2004)

evolutionary dynamics of TSG inactivation. This shows (Fig. 6.) that the number of rate limiting steps to acquiring TSG -/- status is inversely proportional to population size with inactivation occurring with zero rate limiting steps in a large (expanded due to another oncogenic mutation) population. In addition to this, while many TSG mutations are recessive some are dominant negative and therefore mutation of both alleles may not be necessary. For example, p53 binds DNA as a homotetramer. Since the incorporation of a single mutant p53 in this complex disrupts its binding function, even the mutation of a single allele can have a significant effect.

Thus, when several serial rounds of clonal expansion together with the other points mentioned above are taken into consideration, tumorigenesis without a mutator phenotype should at least be considered a possibility

Arguments for the calculations validity

As critics of the mutator phenotype hypothesis have been careful to admit, the above arguments do not deny the possibility of genetic instability being necessary outright. In fact there are several important points which call for the calculation not to be dismissed.

While epidemiological analysis may suggest as few as two mutations could be required, it seems unlikely that they could bring about the six re-wirings thought necessary. As discussed with regard to the previous question, if disruption of TSG's is part of the tumorigenic pathway, two mutations would be necessary for these steps, further increasing the number of mutations required. As the number of mutations required is raised, the probability of cancer developing without increased genetic instability falls geometrically.

Another important fact about epidemiological models is that they rely on clinical data. The intrinsic assumption made is that every cancer cell formed automatically becomes a clinically apparent cancer. However this is unlikely since a cell must evade immune surveillance and successfully induce angiogenesis to grow to the stage at which it is detected clinically. If this is the case, many more cancer cells may arise in a person's lifetime than is evident and a more efficient mechanism of carcinogenesis would be necessary to explain epidemiological data than previously thought.

Tissue Biology

Insights into stem cell and tissue biology reveal the actual level of genetic instability per unit time may be far lower than has been anticipated in any of the models described so far. Tissues are constructed in a way that minimises both variables that normally determine genomic instability: replicative burden and mutation rate per division. The formerly widely accepted epidermal proliferative unit (EPU) model (Lajtha 1979) posits that only a small fraction of the total cell number in a tissue is composed of slowly dividing stem cells (Fig.7.a.).

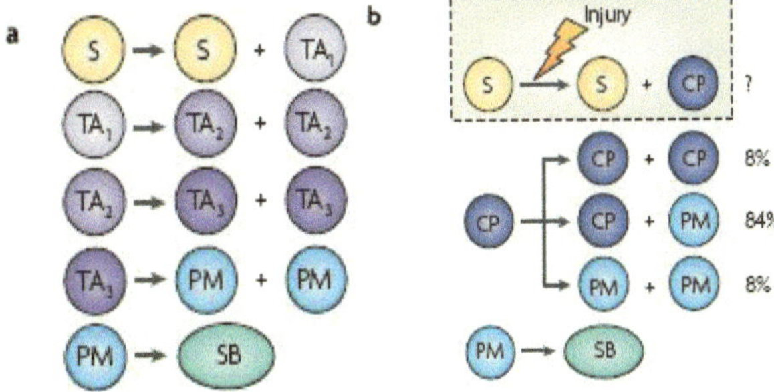

Fig.7.a. The rules of cell fate as dictated by the epidermal proliferative unit (EPU) model52. Self-renewing stem cells (yellow) generate a transit-amplifying (TA) cell population (purple), which undergoes three rounds of symmetric division (TA1 to TA3) before terminal differentiation into post-mitotic (PM) cells (blue). These PM cells subsequently detach from the basal layer and become suprabasal (SB) cells (green). Note that in this model, the behaviour of TA cells is determined by their past history; for example, a TA1 cell will differentiate after two further rounds of cell division. B.The committed progenitor (CP) cell model. On division, daughter CP cells adopt one of three possible fates (the proportion of cells adopting each fate is shown by the percentages). Cells either remain as proliferative CP cells (dark blue) or become terminally differentiating, post-mitotic (PM) basal cells (blue), and subsequently leave the basal layer to become suprabasal cells (SB, green). The box indicates a stem-cell compartment that can generate CP cells, such as those in the hair follicle bulge or interfollicular stem cells. Stem cells (S; yellow) remain quiescent in normal homeostasis but can become activated following injury, dividing to produce CP cells and stem cells. (Figure adapted from Jones and Simons (2008))

This organisation minimises the replicative burden placed on the cells which primarily give rise to tumours. There are also many ways the mutation rate per replication is kept to a minimum. One of the most intriguing mechanisms proposed for minimising the mutation rate is described by the "immortal coil" hypothesis. This posits that one "conserved strand" of DNA from the stem cell is always passed on the daughter stem cell, avoiding copy errors that would otherwise accumulate.

This same hierarchy appears to exist in tumours as well. Al-Hajj et al (2003) found that only 2% of a breast cancer population had tumorigenic potential when injected into NOD or SCID mice and similar results have been reproduced for brain tumours by Singh et al (2003). If this is truly a general phenomenon, these insights suggest that the standard clonal evolution model of cancer is too simplistic. While a clonal expansion does give rise to a greatly expanded population, mutations in progenitors are largely irrelevant, since the evolutionarily important selection only occurs in a miniscule sub-population. The smaller population thus reduces the variation upon which selection can act and the reduced replication burden reduces the rate at which variation is produced. Clearly these changes to the model dramatically increase the predicted time between the steps in the multi-step cancer progression, perhaps beyond a level which can account for actual cancer rates. It therefore suggests that the mutation rate may need to be raised, perhaps several orders of magnitude above that in normal stem cells even to reach the estimates commonly used in the mathematical models. Thus these observations strengthen the case for the mutator phenotype hypothesis which could increase the variation gained by the population per generation and therefore shorten the time between steps.

Based on experiments by Clayton *et al.* (2007) a new "committed progenitor" (CP) model has been developed (Fig.7.b.). The stem cell only divides in the case of injury and is otherwise quiescent. The quiescent state removes the danger of replication errors potentially vastly extending the time delay between steps of the multistep model. In addition to this, calculations show that much as in the EPU model, progenitors are not long lived enough to accumulate sufficient mutations for tumorigenesis before being shed. If this model proves to hold validity in other tissues, it clearly suggests an even greater need for genetic instability than even the EPU model.

However, recent experiments performed by Quintana et al (2008) found that under the right conditions, as many as 27% of single xenografted melanoma cells injected could result in tumour formation suggesting current tumorigenicity tests are underestimating the number of cancer stem cells in cancers. Recent studies including those by Somervaille *et al.* (2009) suggest partially differentiated hematopoietic progenitors can spontaneously revert to an embryonic like transcription program to produce leukemic stem cells. Such new data and the recent development of iPS technology highlight the fact that cell developmental stage and renewal potential may be more malleable than previously thought. This loosens the foundations upon which the stem cell origin model and cancer stem cell hypothesis stand, advocating caution when drawing conclusions based on the previously established dogma in the field of tissue biology.

Epigenetics

Until relatively recently, cancer has been primarily studied as a disease which proceeds via a stochastic accumulation of genetic mutations. In the same spirit, the arguments which have been presented here so far have focused on whether background genetic changes can account for this accumulation. However, cancer is strictly a disease of the heritable deregulation of genes. The genetic changes which have been discussed till now are only one of two major mechanisms which can bring about this deregulation. Normal physiological development shows how gene expression can be drastically altered in a more flexible way, by epigenetic changes. There is a growing body of knowledge (Jones and Baylin 2002; Herman and Baylin 2003; Esteller 2008) that suggests these are also a critical mechanism of gene deregulation in tumorigenesis.

Epigenetic modifications in humans comprise methylation of cytosine 5' to guanine (methCpG) and an array of different combinations of acetylation, methylation and phosphorelation of different histones. The best understood changes are CpG methlations. In normal tissue the great majority of CpG islands are unmethylated, corresponding to loose chromatin and hence an active gene expression state.

Patterns of CpG methylation and chromatin structure are grossly altered during tumorigenesis (Fig.8.), CpG methylation contributes to tumorigenesis via three mechanisms simultaneously:

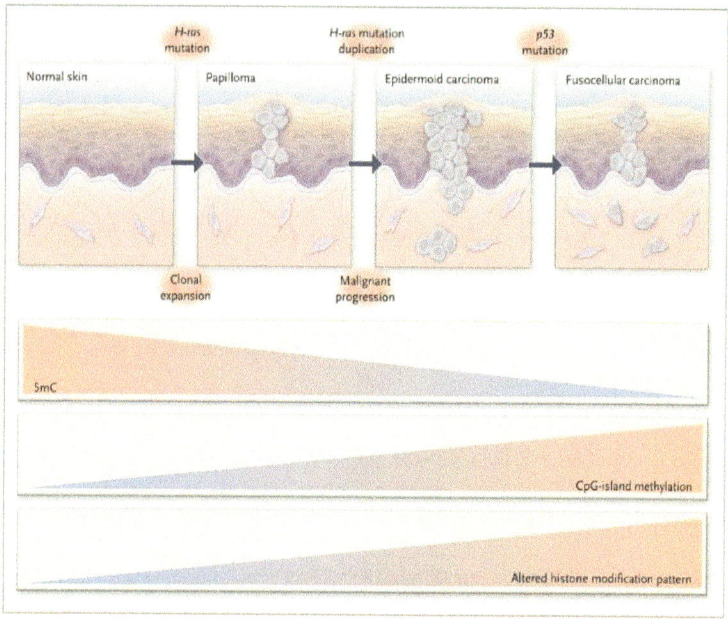

Figure.8. Epigenetic Alterations in Tumour Progression: A multistage model of carcinogenesis in skin is shown. In conjunction with phenotypic cellular changes and the accumulation of genetic defects, there is a progressive loss of total DNA methylation content, an increased frequency of hypermethylated CpG islands, and an increased histone-modification imbalance in the development of the disease. H-ras denotes Harvey–ras oncogene, and 5mC 5-methylcytosine. (Adapted from Estellar 2008)

1. CpG island promoter hypermethylation

2. Global CpG hypomethylation

3. CpG acting as an internal mutagen

CpG island promoter hypermethylation

Silencing of TSG's by promoter methylation has been shown to occur in almost all

cancers examined to date (Baylin and Herman 2000; Jones and Laird 1999). The fact

that as many, if not more genes are inactivated by hypermethylation as by

mutations, highlights TSG inactivation as a major event in tumour progression (Fig.

9.).

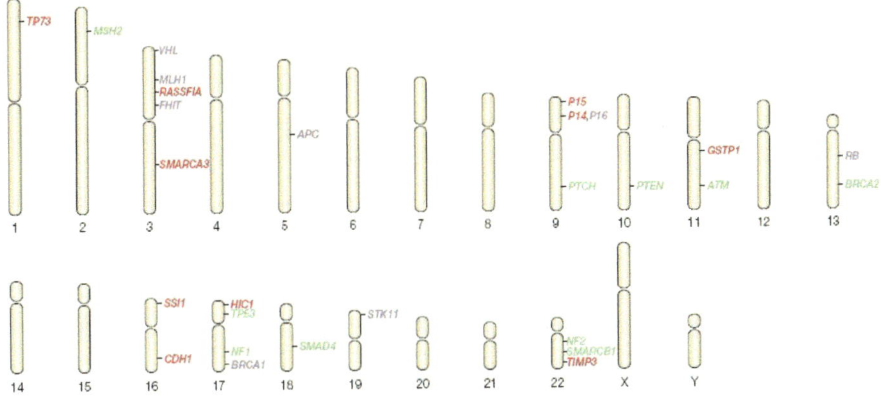

Figure .9. A map of the human genome: Genes that are frequently mutated and/or hypermethylated in cancer. Only a subset of the genes that are known to be frequently hypermethylated and silenced in one or more cancer types, and genes that are most often mutated in tumours are shown here. Nearly all of these genes are in chromosome regions that commonly show loss of heterozygosity in cancer. Genes for which only genetic mutations have been reported are shown in green, those that have been reported to be only hypermethylated are shown in red and those for which both changes have been reported are shown in purple. In the 9p region, the individual transcripts *P16INK4A* and *P14ARF* are transcribed from the *CDKN2A* locus, and are each shown because of the different incidences of hypermethylation and mutation at their promoters. It is noteworthy that as many, if not more, genes are inactivated by promoter hypermethylation and by epigenetic silencing as they are by coding-region mutations. *ATM*, ataxia elangiectasia mutated; *APC*, adenomatosis polyposis coli; BRCA1/2, breast cancer1/2, early onset; *CDH1*, E-cadherin; *CDKN2A/B*, cyclin-dependent kinase inhibitor 2A/B; *FHIT*, fragile histidine triad; *GSTP1*, glutathione S-transferase pi; *MLH1*, mutL homologue 1, colon cancer, non-polyposis type 2; *MSH2*, mutS homologue 2, colon cancer, non-polyposis type 1; *NF1/2*, neurofibromin 1/2; *PTCH*, patched homologue; *PTEN*, phosphatase and tensin homologue; *RB1*, retinoblastoma 1; *SMAD4*, mothers against decapentaplegic homologue 4; *SMARCA3/B1*, SWI/SNF-related, matrix-associated, actin-dependent regulator of chromatin, subfamily A, member 3/subfamily B, member 1; *STK11*, serine/threonine kinase 11; *TIMP3*, tissue inhibitor ofmetalloproteinase 3; *TP53/73*, tumour protein p53/p73; *VHL*, von Hippel–Lindau syndrome. (Reproduced from Jones and Baylin 2002)

This is can clearly have a critical effect since it can provide the second "hit"

required to disrupt most TSG function e.g. in patients with sporadic breast cancers, the first BRCA1 allele is either mutated or removed by a LOH event and in 10-15%

the second is hypermethylated (Esteller 2000).

Numerous studies of the colorectal cancer (CRC) model have shown methCpG

mediated silencing can play roles in both the initiation and later progression of

cancer. Issa *et al.* (1999) have shown there is a progressive age related methylation

of a specific subset of genes in normal mucosa (Table.1. reproduced from Issa. 2000).

TABLE 1. Genes affected by age-related promoter methylation in normal colon and colorectal cancer

Gene	Map	Expressed in normal colon?	Methylation in colon cancer (%)
Promoters methylated in aging colon and colon cancer			
CSPG2 (*Versican*)	5q12–14	yes	70%
EGFR	7p12	yes	0%
ER (estrogen receptor)	6q25.1	yes	>90%
IGF2 (P2-4)	11p15.5	yes (low levels)	70%
MYOD1	11p15.4	no	>90%
N33	8p22	yes	80%
PAX6	11p13	?	70%
RARβ2	3p24	yes	10%
Promoters methylated in colon cancer only			
APC	5q21	yes	10–20%
CACNA1G	17q22	yes	10–20%
CALCA	11p15	no	50%
HIC1	17p13.3	yes (low levels)	50%
hMLH1	3p21.3	yes	10–20%
^{6}O-MGMT	10q26	yes	30%
p14/ARF	9p21	yes	10–20%
p16 (CDKN2A)	9p21	yes	20–30%
THBS1	15q15	yes	10–20%
TIMP3	22q12–13	yes	27%
WT1	11p13	?	50%

Since these studies have only covered a small fraction of the genome it is reasonable

to assume there are many other genes affected in this way. As is to be expected,

such wide ranging changes have been shown to result in a substantial expansion of

the proliferative zone also commonly seen in the otherwise normal epithelium of colorectal cancer patients (Roncucci *et al.* 1988). In support of methylations' causal influence on colorectal cancer tumorigenesis, mice with pharmacologically lowered age-related methylation had a substantial reduction in proliferative zone expansion and tumour incidence (Laird *et al.* 1995).

A different subset of genes has been shown to be silenced exclusively in neoplasia (Table.1.). These appear to result from a so called CpG island methylator phenotype (CIMP). In fact ~50% of unselected CRC cases have been found to be CIMP+ (Toyota *et al.* (2000)). This could provide an alternative mechanism to genetic instability for producing variation in tumour progression (Fig.10.). Although not universally accepted as individual pathways, they have their own distinctive sets of mutations (Fig.10.). It is an oversimplification however to suggest these are entirely independent pathways. There is a great deal of overlap e.g. CIMP+ tumours can display MIN usually due to hMLH silencing.

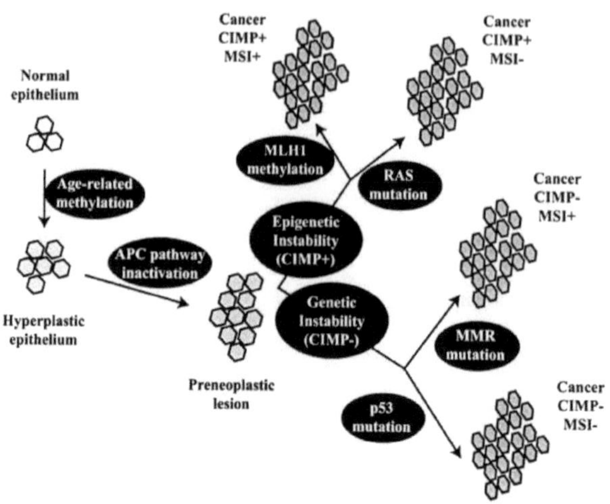

FIGURE 10. A model of the molecular aetiology of colorectal cancer. In this model, the colorectal epithelium starts out normal (*left*), and becomes hyperplasic as a result of age related events, including CpG island methylation. The hyperplasic "at risk" mucosa then acquires a genetic defect in the APC pathway and forms a preneoplastic lesion. Progression from this lesion follows two major pathways: epigenetic (CIMP+, *top*), or genetic (CIMP–, *bottom*). In each group, cancers develop along either an MSI+ pathway if mismatch repair (MMR) genes are inactivated, or an MSI– pathway, characterized by Ki-*ras* mutations in CIMP+ cases, and *p53* mutations and chromosomal instability in CIMP– cases. (Reproduced from Issa. 2000)

Global CpG hypomethylation

The few genes that are normally silenced in cells are developmental stage, tissue specific genes and those silenced by genomic imprinting. Loss of imprinting (LOI) can therefore lead to biallelic expression of oncogenes. A well studied example is the LOI of Insulin-like growth factor 2 (IGF2) which leads to a modest increase in expression and is found in the colonic mucosa of 30% of colorectal cancer patients. Sakatani *et al.* 2005 produced a mouse model of this and found that mice with LOI had a 2 to 2.5 fold increased risk of tumours when compared with their control littermates. This effect may not only be due to the modest increase in IGF2 expression. Both in the human and mouse the intestinal epithelium showed a shift toward less differentiated epithelium. A general loss of differentiation can channel stem cells into abnormal clonal expansion, in fact Holm *et al.* 2005 have shown that global LOI can immortalise cells in vitro when cooperating with ras mutations. This has led them to suggest LOI and silencing events may be sufficient for tumorigenesis initiation. The clearest demonstration of this is that embryonic stem cells injected into immunocompromised mice can form teratomas despite having a pristine genome.

The changes described above provide a mechanism in parallel to genetic mutation by which TSG's can be knocked out and oncogenes activated. In doing so epigenetic changes take a significant amount of the onus off genetic change in explaining tumorigenesis and therefore weaken the case significantly for the mutator phenotype being strictly necessary.

Does Genetic instability accelerate tumour progression?

Loeb has made it clear that the mutator phenotype hypothesis and clonal evolution are by no means mutually exclusive; in fact, it is likely that both hypotheses contribute to tumour progression in parallel. While it has been shown above that the mutator phenotype may not be strictly necessary theoretically, our understanding of cancer and our ability to treat and prevent it depend critically on knowledge of the mechanisms and pathways exploited by nature in practice.

Cell clone ecology hypothesis

The somatic tissues of metazoans can be viewed as populations of asexually reproducing organisms which are undergoing a lifelong struggle with their neighbours for increased replication. Immediate fitness of the individual e.g. due to acquiring any of the six classic hallmarks, is not the only heritable trait. Other traits which can be selected for include those which influence the entire lineages' fitness, although selection pressures for these will always be subordinate to those with immediate effects. What follows is a mathematical assessment of whether the mutator phenotype, being such a trait, is beneficial to tumorigenesis and thus might be expected to play a role in human cancer.

Mathematical assessment

1. Increase in deleterious mutation load

The potentially most damaging criticism of the mutator phenotype hypothesis is that reduction in fitness due to deleterious mutations might outweigh the benefit of the small fraction of beneficial mutations. There is clearly a limit beyond which any further increase in genetic instability leads to a greater accumulation of deleterious "reduced fitness" (RF) mutations than the cell can remain viable with and negative clonal selection (NCS) leads to extinction of these clones

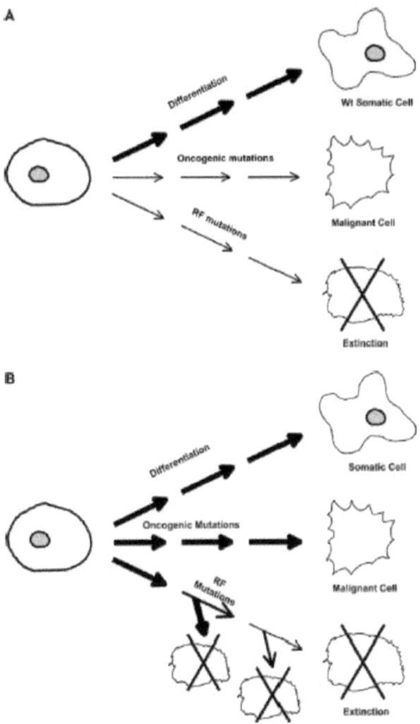

Figure 11.—Fates of cell lineages without (A) and with (B) initial mutator mutations. The major pathway for normal stem cells (A) is to differentiate into somatic cells, but development of oncogenic mutations leading to cancer or of mutations leading to reduced fitness and clonal extinction also occurs at a low frequency. The introduction of a mutator mutation (B) results in increased frequency of oncogenic mutations and increased frequency of mutations leading to reduced fitness. (Reproduced from Loeb et al. 2005)

(Fig .11.). However Loeb et al (2005) considered NCS in isolation (increased fitness mutations which raise mutator clone fitness are disregarded), apoptosis evading mutations (which would reduce NCS) are disregarded and any mutation which creates even the slightest reduction in fitness results in immediate elimination of the clone. Such stringent assumptions ensure the model can be considered a limiting case , despite this, the model predicts that NCS does not limit growth of premalignant clones significantly in the great majority of cases examined.

2. Competition with random background mutants

Given the insight from the calculation above, Beckman and Loeb (2006) have

performed a mathematical analysis of the relative efficiency of carcinogenesis with

and without a mutator mutation. By calculating the relative efficiency they are able

to sidestep the question of whether genetic instability is necessary and focus on the

question of whether genetic instability is a beneficial trait overall.

Figure.12. below shows the relative expected contribution of mutator versus

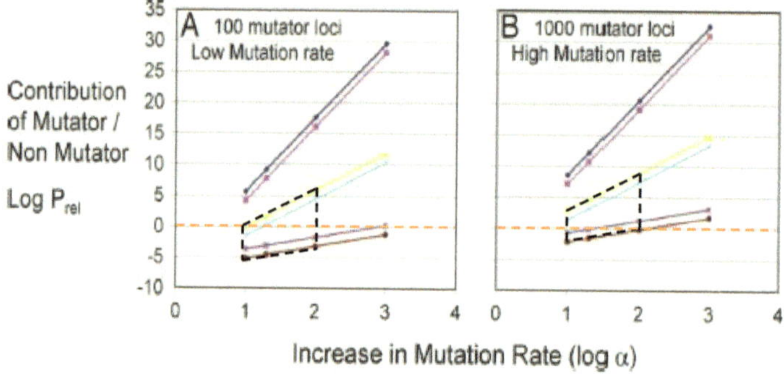

Fig. 12. Log(Prel, mut:no mut), the common logarithm of the relative efficiency of carcinogenesis with a mutator mutation compared to that without a mutator mutation, plotted vs. log α, the common logarithm of the fold increase in mutation rate due to a mutator mutation. Assuming the number of nucleotide loci available for a mutator mutation N_{ML} is 100 (A) or 1,000 (B) and that the initial mutation rate k_{mut} is 10^-11 (A) or 10^-9 (B). T (the number of cell generations at which the relative efficiency of different mechanisms is evaluated) either 170 (brown, light blue, and magenta) or 5,000 (purple, yellow, and dark blue). C (the number of oncogenic mutations required to produce a cancer) 2 (lower lines), 6 (middle lines), or 12 (upper lines). (Adapted from Beckman and Loeb. 2006)

non-mutator clones to clinical cancers, assuming a cell will form a clinically apparent

cancer once it has attained sufficient mutations. An initial glance reveals most of the

lines are above the level of zero relative efficiency and therefore that the results

indicate mutator clones will outcompete non-mutators in most cases. However a

stringent assessment of what is likely the case in the majority of cancers may be

more revealing. Both graphs should be considered since evidence collected so far

(Myung and Kolodner 2002) suggests there could be 100 - 1000 or more dominant

mutator loci. Background mutation rates cannot currently be fixed (see stem cell discussion above) but are likely to be between 10^{-11} and 10^{-9} per nucleotide per generation. To make this discussion as widely applicable as possible assessment at both 170 and 5000 generations should be considered. This takes into account the range of different stem cell models rates of tissue cycling (e.g. slow cycling brain and fast cycling colon). In another paper Loeb and Beckman (1993) show that mutator mutations in proofreading can increase mutation rate by between 10 and 1000 fold. However most observed increases have been between 10 and 100 so for the most stringent assessment this range will be considered (α=10-100).

When six mutations are required (as discussed previously) then given all of the previous considerations the area between the light blue line and the black dashed line should be considered. According to this model since most of this area is above zero (orange dashed line) on the graph, mutator pathways are relatively more efficient in the great majority of cases.

However one must also consider the case where only 2 oncogenic mutations are required, it appears the opportunity cost of an immediate fitness advantage given by an oncogenic mutation outweighs the benefit of mutator mutations. Until the relatively recently this would only have been thought of as occurring in exceptional cases e.g. some paediatric cancers. However this could have more widespread relevance if it is considered that epigenetic silencing and LOI could reduce the required number of mutations below the originally postulated six. Thus, epigenetic considerations increase the uncertainty to the areas outlined by the black dashed lines as possibly being more representative of the mutator advantage in actual carcinogenesis.

Lab based test

Confirmation that high mutation rates can evolve under the correct selection pressures comes from laboratory scale evolution experiments. Lac⁻ *E.coli* were put through selection procedures for the Lac⁺ allele (Mao *et al.* 1996). Mis-match repair (MMR, resulting in MIN-type genetic instability) deficient strains increased in relative frequency from $1{:}1{\times}10^{5}$ to $1{:}2{\times}10^{2}$. A wide variety of lab scale evolution experiments (Shaver *et al.* 2002; Labat *et al.* 2005) have confirmed these results, not only showing that mutator clones reach fixation, but also that they do so because of the increased genetic instability rather than due to any direct benefit (see Bardelli criticism later).

Clinical data

Since there are numerous different pathways to cancer and assuming these are all competing simultaneously, it is likely tumour progression mechanisms will manifest themselves in a ratio according to their relative efficiencies. Therefore, while not proving a causal relationship, a widespread incidence of the mutator phenotype in the population of clinical cancers certainly strengthens the case for the mutator mutations accelerating effects in nature.

In accord with the models and experiments above, it is widely accepted that most cancers do display some sort of genetic instability. As a result of the many new sets of genes which have been discovered, the mutator phenotype hypothesis been expanded from postulating just errors in DNA polymerases to include: microsatellite instability (MIN), chromosomal instability (CIN), checkpoint instability and genes involved in epigenome maintenance (Loeb *et al.* 2008). Each of these types of instability is represented in many different types of cancer.

A nearly ubiquitous type of genetic instability is CIN (Loeb. 2003). In many major types of cancer this can regularly lead to allelic losses of up to 50%, allowing relatively rapid deletion of important TSG's. A very high percentage of colorectal tumors exhibit CIN (with the rest showing MIN), contrary to what the critics of Loeb would speculate, Vogelstein et al (1997) have shown that this is not merely an end product of increased cycling during tumorigenesis. Compared to normal cells undergoing the same number of divisions in the same medium, CIN colorectal cancers exhibited an increase of 10^{-2} chromosome number changes per generation.

At the very least, in the case of colorectal cancer the high incidence of genetic instability confirms theoretical predictions to suggest that cancer progression is accelerated.

While such clinical data may lend credence to the mutator model it does not confirm a causal relationship. Some authors (Brevik. 2001) have suggested, that since the mutator phenotype does not itself provide any immediate fitness gain it may not be genetic instability *per se* that accelerates cancer. In normal environments it may be beneficial to stop the cycle for DNA repair, however, in the environments where cancers often evolve, such as the mutagen coated epithelium of smokers lungs, it could be a disadvantage. It would clearly provide an immediate fitness gain if the deregulation in repair checkpoints accompanying genetic instability allows the cell to continue cycling while normal cells are frequently arrested in G1. Indeed this selection pressure has recently been shown to be important in vitro by Bardelli *et al*; different types of genetic instability were selected for depending on the presence of otherwise cytotoxic levels of mutagens.

The strong selection pressure exerted by mutagens in this way is further illustrated by their effects in some other experiments by Mao *et al.* (1996). They showed a mutagen round of selection increased the proportion of mutators to non-mutators to 1:2 or 1:1 (with two or more rounds) rather than to 1:200 seen when only selection for Lac+ mutants was performed (as discussed above). Thus many cases of genetic instability may be explained as an adaption to adverse conditions rather than being due to the necessity of the mutator phenotype for tumorigenesis.

Sequence Data

Stoler *et al.* (1999) reported that the average cancer has >11000 genetic alterations per cell. Other authors have found many more and due to the inefficiency of detection these are likely only a fraction of the actual number. Such high values, were greater than both the authors themselves, and many commentators (Boland and Ricciardiello 1999) had expected. This and the discovery that early stage adenomatous polyps also harboured considerable numbers of genetic alterations was taken by some as evidence in favour of genetic instability having a crucial causal role in tumorigenesis.

However according to simple mathematical models of colon cancer (below) produced by Tomlinson and Bodmer (2002) this number of mutations can be accounted for more than adequately without invoking the need for genetic instability or even clonal evolution and expansion. According to a conservative estimate based on this model, all tumors should be expected to accumulate 10^{12} mutations. This large number is mainly due to the large number of generations a stem cell is assumed to cycle through due to the large size of an end stage tumour and the high rate of cell death in the tumour environment.

The assumptions of the number of cycles a stem cell is forced to go through during tumorigenesis could be considered grossly overestimated, especially when the CP model (discussed in the previous section) is considered. It has also been pointed out (Loeb. 2008) that a mutator mutation may be required just to reach the mutation rate used in this calculation. Despite such criticism the model does suggest mutations are likely far more common than many expected, weakening the standing of the sequence data as well as some of the clinical data as evidence for the mutator hypothesis.

Implications for therapy

The insights into the mechanisms of cancer evolution have implications for cancer therapy. While genetic instability may accelerate tumor progression it may also be its Achilles heel. The effects of NCS have been shown to be minimal in most cases (above) but the tumor cells are clearly closer to the threshold where NCS could have an effect. There is therefore a basis for a good therapeutic index for drugs which raise genetic instability (in fact many current chemotherapeutics can) beyond this threshold in cancer cells but not normal cells. Conversely, Knowledge that genetic instability accelerates progression suggests lowering genetic instability may slow carcinogenesis. Thus in this case, understanding of the fundamental mechanisms of tumorigenesis may potentially lead to a form of "prevention by delay" (Beckman and Loeb 2005) i.e. delay of disease onset till a time late in or past the normal lifespan.

Conclusion

This analysis has highlighted the fact that genetic instability is not strictly necessary for human cancer. However it has also made it clear that it can increase the efficiency of carcinogenesis substantially in many cases, as demonstrated by the fact that it occurs in the majority of cancers. This suggests it is an important factor in

tumour progression and is thus worthy of future research into its basis and function.

Importantly, the insights from this discussion also suggest a potentially distinct set of

potential drug targets for both cancer prevention and therapy.

Bibliography

Al-Hajj, M., M. S. Wicha, et al. (2003). "Prospective identification of tumorigenic breast cancer cells." Proc Natl Acad Sci U S A **100**(7): 3983-8.

Armitage, P. (1985). "Multistage models of carcinogenesis." Environ Health Perspect **63**: 195-201.

Beckman, R. A. and L. A. Loeb (1993). "Multi-stage proofreading in DNA replication." Q Rev Biophys **26**(3): 225-331.

Beckman, R. A. and L. A. Loeb (2005). "Genetic instability in cancer: theory and experiment." Semin Cancer Biol **15**(6): 423-35.

Beckman, R. A. and L. A. Loeb (2005). "Negative clonal selection in tumor evolution." Genetics **171**(4): 2123-31.

Beckman, R. A. and L. A. Loeb (2006). "Efficiency of carcinogenesis with and without a mutator mutation." Proc Natl Acad Sci U S A **103**(38): 14140-5.

Bodmer, W., J. H. Bielas, et al. (2008). "Genetic instability is not a requirement for tumor development." Cancer Res **68**(10): 3558-60; discussion 3560-1.

Boland, C. R. and L. Ricciardiello (1999). "How many mutations does it take to make a tumor?" Proc Natl Acad Sci U S A **96**(26): 14675-7.

Clayton, E., D. P. Doupe, et al. (2007). "A single type of progenitor cell maintains normal epidermis." Nature **446**(7132): 185-9.

Cook, P. J., R. Doll, et al. (1969). "A mathematical model for the age distribution of cancer in man." Int J Cancer **4**(1): 93-112.

Dong, Y., M. A. Hakimi, et al. (2003). "Regulation of BRCC, a holoenzyme complex containing BRCA1 and BRCA2, by a signalosome-like subunit and its role in DNA repair." Mol Cell **12**(5): 1087-99.

Esteller, M. (2000). "Epigenetic lesions causing genetic lesions in human cancer: promoter hypermethylation of DNA repair genes." Eur J Cancer **36**(18): 2294-300.

Esteller, M. (2008). "Epigenetics in cancer." N Engl J Med **358**(11): 1148-59.

Esteller, M., J. M. Silva, et al. (2000). "Promoter hypermethylation and BRCA1 inactivation in sporadic breast and ovarian tumors." J Natl Cancer Inst **92**(7): 564-9.

Fearon, E. R. and B. Vogelstein (1990). "A genetic model for colorectal tumorigenesis." Cell **61**(5): 759-67.

Fisher RA. "The Genetical Theory of Natural Selection". 2nd ed. New York: Dover (1st Ed Oxford Univ Press); 1930.

Foulds, L. (1954). "The experimental study of tumor progression: a review." Cancer Res **14**(5): 327-39.

Greenman, C., P. Stephens, et al. (2007). "Patterns of somatic mutation in human cancer genomes." Nature **446**(7132): 153-8.

Hahn, W. C., C. M. Counter, et al. (1999). "Creation of human tumour cells with defined genetic elements." Nature **400**(6743): 464-8.

Hanahan, D. and R. A. Weinberg (2000). "The hallmarks of cancer." Cell **100**(1): 57-70.

Herman, J. G. and S. B. Baylin (2000). "Promoter-region hypermethylation and gene silencing in human cancer." Curr Top Microbiol Immunol **249**: 35-54.

Herman, J. G. and S. B. Baylin (2003). "Gene silencing in cancer in association with promoter hypermethylation." N Engl J Med **349**(21): 2042-54.

Issa, J. P. (1999). "Aging, DNA methylation and cancer." Crit Rev Oncol Hematol **32**(1): 31-43.

Issa, J. P. (2000). "The epigenetics of colorectal cancer." Ann N Y Acad Sci **910**: 140-53; discussion 153-5.

Jones, P. and B. D. Simons (2008). "Epidermal homeostasis: do committed progenitors work while stem cells sleep?" Nat Rev Mol Cell Biol **9**(1): 82-8.

Jones, P. A. and S. B. Baylin (2002). "The fundamental role of epigenetic events in cancer." Nat Rev Genet **3**(6): 415-28.

Jones, P. A. and P. W. Laird (1999). "Cancer epigenetics comes of age." Nat Genet **21**(2): 163-7.

Kolodner, R. D., C. D. Putnam, et al. (2002). "Maintenance of genome stability in Saccharomyces cerevisiae." Science **297**(5581): 552-7.

Labat, F., O. Pradillon, et al. (2005). "Mutator phenotype confers advantage in Escherichia coli chronic urinary tract infection pathogenesis." FEMS Immunol Med Microbiol **44**(3): 317-21.

Laird, P. W., L. Jackson-Grusby, et al. (1995). "Suppression of intestinal neoplasia by DNA hypomethylation." Cell **81**(2): 197-205.

Lajtha, L. G. (1979). "Stem cell concepts." Differentiation **14**(1-2): 23-34.

Lengauer, C., K. W. Kinzler, et al. (1997). "Genetic instability in colorectal cancers." Nature **386**(6625): 623-7.

Loeb, L. A. (1991). "Mutator phenotype may be required for multistage carcinogenesis." Cancer Res **51**(12): 3075-9.

Loeb, L. A., J. H. Bielas, et al. (2008). "Cancers exhibit a mutator phenotype: clinical implications." Cancer Res **68**(10): 3551-7; discussion 3557.

Loeb, L. A., K. R. Loeb, et al. (2003). "Multiple mutations and cancer." Proc Natl Acad Sci U S A **100**(3): 776-81.

Luebeck, E. G. and S. H. Moolgavkar (2002). "Multistage carcinogenesis and the incidence of colorectal cancer." Proc Natl Acad Sci U S A **99**(23): 15095-100.

Mao, E. F., L. Lane, et al. (1997). "Proliferation of mutators in A cell population." J Bacteriol **179**(2): 417-22.

Moolgavkar, S. H. and A. G. Knudson, Jr. (1981). "Mutation and cancer: a model for human carcinogenesis." J Natl Cancer Inst **66**(6): 1037-52.

Nowak, M. A., F. Michor, et al. (2004). "Evolutionary dynamics of tumor suppressor gene inactivation." Proc Natl Acad Sci U S A **101**(29): 10635-8.

Quintana, E., M. Shackleton, et al. (2008). "Efficient tumour formation by single human melanoma cells." Nature **456**(7222): 593-8.

Rangarajan, A., S. J. Hong, et al. (2004). "Species- and cell type-specific requirements for cellular transformation." Cancer Cell **6**(2): 171-83.

Roncucci, L., M. Ponz de Leon, et al. (1988). "The influence of age on colonic epithelial cell proliferation." Cancer **62**(11): 2373-7.

Shaver, A. C., P. G. Dombrowski, et al. (2002). "Fitness evolution and the rise
    of mutator alleles in experimental Escherichia coli populations."
    <u>Genetics</u> **162**(2): 557-66.
Somervaille, T. C., C. J. Matheny, et al. (2009). "Hierarchical maintenance of
    MLL myeloid leukemia stem cells employs a transcriptional program
    shared with embryonic rather than adult stem cells." <u>Cell Stem Cell</u>
    4(2): 129-40.
Stein, W. D. and A. D. Stein (1990). "Testing and characterizing the two-stage
    model of carcinogenesis for a wide range of human cancers." <u>J Theor
    Biol</u> **145**(1): 95-122.
Tomlinson, I. P., M. R. Novelli, et al. (1996). "The mutation rate and cancer."
    <u>Proc Natl Acad Sci U S A</u> **93**(25): 14800-3.
Toyota, M., F. Itoh, et al. (2002). "DNA methylation changes in gastrointestinal
    disease." <u>J Gastroenterol</u> **37 Suppl 14**: 97-101.
Toyota, M., M. Ohe-Toyota, et al. (2000). "Distinct genetic profiles in
    colorectal tumors with or without the CpG island methylator
    phenotype." <u>Proc Natl Acad Sci U S A</u> **97**(2): 710-5.
Weinberg, R. (2007) "The biology of cancer". <u>Garland Science</u>